Treasures of the Sea

Treasures of the Sea

Marine Life of the Pacific Northwest

by James Cribb

Toronto
OXFORD UNIVERSITY PRESS
1983

This book is dedicated to my wife
ANDREA

ACKNOWLEDGEMENTS

I wish to thank the following people, whose efforts
have helped to bring about this book—
Flo and Trevor Anderson, Mark Atherton,
Jim Borrowman, Brian Congdon, Andrea Cribb,
Ralph Delisle, Peter Dickinson, Donna and Bill
Mackay, Bill McBay, Bob McBay, Val Ranetkins,
Gina and Vaughn Raymond, Mike Richmond,
and Roger Boulton. My special thanks are
due to Dr David O. Duggins, of Friday Harbor
Laboratories, University of Washington, for editing
the captions and verifying their accuracy.

CANADIAN CATALOGUING IN PUBLICATION DATA

Cribb, James, 1956–
   Treasures of the sea

Includes index.
ISBN 0-19-540418-1

1. Marine biology—Northwest coast of North America.
2. Marine fauna—Northwest coast of North America.
I. Title.

QH95.3.C75   574.97'000964'3   C82-095336-9

Produced by Roger Boulton
Designed by Fortunato Aglialoro

Printed in Hong Kong by
EVERBEST PRINTING COMPANY LIMITED

ISBN 0-19-540418-1

1 2 3 4 - 6 5 4 3

Introduction

'. . . none shall contemplate anything
whatsoever but that he shall see God therein.'
Bahá'u'lláh: The Seven Valleys

Under the tides of the Pacific northwest lies hidden a mysterious world of wildlife without parallel in its strange variety. Veiled by the current-swept and often murky waters of the ocean is an array of stunningly beautiful, intricately detailed animals, whose flower-like delicacy calls to mind the gentle slopes of an alpine meadow, rather than the frigid depths of a hostile sea. Save for a handful of divers who make brief incursions into this environment, or those who catch a glimpse of its life in aquaria on land, few people anywhere know that such a unique congregation of animals and plants exists. The sea guards its secrets well.

Five years ago I put on scuba gear and dropped down into this world beneath the waves. As I drifted weightlessly over the ocean floor, my eyes were saturated with colours, shapes, patterns, and forms such as I had never seen before. Our senses of smell, touch, taste and hearing are almost useless underwater. Only by vision can we perceive. Where eyes are the only tool of the mind, vision is sharpened, the power of sight enhanced. Nothing else can rival the brilliance of what we then see.

There lay before me a multitude of exciting flora and fauna that few could have even imagined. It seemed that such an amazing work of creation must not go unrecorded, and photography was the obvious medium to use. Only the camera could portray these subjects in their natural surroundings. I spent four years within those waters. This book is a footnote to that experience, simply a brief review of one of the greatest shows on earth.

People often ask me what it is like to take pictures underwater and how it is done. Using 35mm. cameras and lenses, mounted in protective housings of metals and plastics, I can manipulate most controls with levers and rods. One cannot change film or lenses underwater, so I usually take three cameras on a dive, each one equipped with a different lens, and leave two of them hanging from the boat, readily accessible in case I finish a roll of film or need another lens. Over the past ten years equipment for this kind of work has much improved in design and dependability, but even so mechanical failures are all too frequent. While I was filming the diver with the octopus (plate 62) a leak developed in the camera housing. Encounters with these giant molluscs are rare, so I chose to stay and risk the camera, rather than lose a chance that I might never have again. Fortunately, this time, I was able to finish my shooting with nothing worse than moisture on the lens.

Time is precious underwater. The length of a dive is limited by many factors: the amount of compressed air carried by the diver, the depth, as times and depths must be strictly observed to prevent decompression sickness, the rapid chilling effect of the water. One of the most important considerations in the timing of a dive is the danger of sudden changes in the ocean currents. The animals of the seabed thrive best in areas where the currents sweep vast amounts of food within their reach and therefore photographers seek out such places for their abundant life, but currents can be swift: it may be that the only safe time in the water is the interval when they slacken with the change of tide. The sculpin and anemone (plate 7) were photographed in Seymour Narrows, off Vancouver Island, a channel infamous for unpredictable currents reaching speeds of fourteen and fifteen knots and swinging around so quickly that a fifteen-minute slack time is a blessing. I spotted the sculpin after the tide had changed and as the current was getting stronger. Good sense would have had me leave the water, but the appeal of this little grouping was too much to resist. I braced myself therefore, took the photograph, finished the roll of film and then began drifting back to the boat. Suddenly I was trapped in a violent current that forced me down fifty feet before I could grab hold of an outcrop of rock to stop my descent, and I regained the surface only by scaling the rock-face, literally hand over hand.

Decompression sickness, commonly known as 'the bends', is debilitating and can be deadly. To avoid it, a scuba diver must work to a fixed set of tables which dictate the tolerable span of time for any given depth. For example, the fish in the cloud sponge (plate 67) was photographed thirty-six metres (one hundred and twenty feet) down, a depth at which one can stay for only fifteen minutes before ascending. Since ocean wildlife is unpredictable and elusive, it may take many fruitless dives before one finds the sought-after pose; such was the case with this particular photograph.

Often people ask me about dangerous sea creatures. I am happy to say there are few. The giant red sea-urchin (plate 36) probably 'injures' more divers than any other animal, but the sea-urchin is not to blame if a human being comes along and kneels on it.

Dangerous sharks are not common, but Pacific dogfish are. Individually, these sharks present no problem to the diver, but when they pack together their annoying habit of coming at one and bumping with their noses can be unnerving. The dogfish in plate 53 was but one of ten or eleven swimming around, and soon after taking that picture I left the water because I could no longer keep a wary eye on them all.

The *Cyanea* jellyfish can inflict a painful sting, as I found in taking its picture (plate 52). To emphasize the jellyfish itself, I had moved in close with a wide-angle lens, and a few of the trailing tentacles brushed across my face. However, the resulting image lasts long after the burning red welts are forgotten.

Subtidal animals are fiercely competitive, endlessly striving against the odds, to exist, to feed, to multiply so that the species may endure. To man, the crimson anemone's tentacles may be beautiful or marvellous. To the helpless sea creature that blunders into their paralyzing embrace, these lovely tentacles are living death.

The undersea world assuredly holds its perils, for conditions can be extremely inhospitable. Yet, in some paradoxical way, it leaves tranquillity in the soul. Often I have wondered at the reason for so much beauty, where there is none to see it and so little light by which to see. I have tried to suggest something of this mystery in my work.

As a photographer, I use words sparingly, choosing rather to express myself in images. Therefore the captions in this book are brief. Moreover, knowledge of the differences between subtidal species is as yet largely confined to biological characteristics resulting from such technicalities as current action, water temperature, salinity, variations of habitat, substrata, and the like. Such an amount of detail, it seemed to me, would not have been appropriate to this book and for the same reason I chose not to overload the captions with the Latin nomenclature.

All through the following pages are creatures whose ancestors predate man by millions of years. Slowly, almost imperceptibly, they are being subjected to the crushing weight of man's indifference, his disregard of their natural right to exist. Millions of gallons of pollutants are poured into their waters every year; over-fishing and over-hunting destroy the food-chains on which these forms of life depend; and oil-tankers pose an ever-present threat of sudden extinction. The waters of the Pacific northwest contain an incalculable treasure; I pray it will not be destroyed in the name of progress.

A Note on the Captions and the Photographic Information

Captions descriptive of the subjects have been set slightly apart from the plates themselves. When viewing a photograph, I find that adjacent information can be distracting. Therefore descriptions are grouped in advance of every dozen plates.

The technical aspects of underwater photography cannot be ignored. Shutter speeds and aperture sizes, focal lengths and film types, choice of camera equipment, all affect such qualities as perspective, texture, and colour reproduction. The technical information for each photograph will be found on pages 121 to 126.

All photographs, except plate 70, were taken with a Nikon F or F2 camera body equipped with a DA-1 action finder. Supplementary lighting was provided by Oceanic 2001 submersible strobes. When making so many scores of references to photographic equipment bearing patented or registered trade names, it has simply not been possible to mark each one as copyright or registered in the body of the text. I therefore apologize in advance to the manufacturers whose excellent materials and equipment I have used that I have not been able to indicate trademark registration item by item.

PLATE 1. _A scuba diver hovers above a bed of bull kelp. These plants grow to over 30.5 metres (100 feet) in length, forming dense forests that provide shelter and food to many undersea animals._

Depth: 2.4 m. (8 ft)

PLATE 2. _A scuba diver's light sets aglow a wall swarming with life. Water filters out most of the sun's rays, so that a light source is required to reveal the intrinsic colours._

Depth: 9.1 m. (30 ft)

PLATE 3. _A crimson anemone clings to an underwater precipice, splaying its tentacles to ensnare tiny organisms. Kelp is seen in the background._

Diameter (anemone): 20.3 cm. (8 in.)
Depth: 6.1 m. (20 ft)

PLATE 4. _A swimming anemone bends in sympathy with a swift current. When alarmed, this anemone will release its hold and swim to a new location._

Height: 17.8 cm. (7 in.)
Depth: 7.6 m. (25 ft)

PLATE 5. _Silver-tipped anemones. While anemones are actually animals, they often resemble flowering plants._

Height: 7.6 cm. (3 in.)
Depth: 12.2 m. (40 ft)

PLATE 6. _Sculpin. This is a small member of the sculpin family of fish, most of which are sluggish bottom-dwellers._

Length: 6.4 cm. (2¹/₂ in.)
Depth: 7.6 m. (25 ft)

PLATE 7. _A red Irish lord, a common member of the sculpin family, rests on the bottom beside a closed sea anemone._

Length: 25.4 cm. (10 in.)
Depth: 9.1 m. (30 ft)

PLATE 8. _Two_ Tealia _anemones. Widespread through the Pacific northwest in a variety of colours,_ Tealia _anemones often bear striped patterns over most of their bodies._

Diameter: 10.2–12.7 cm. (4–5 in.)
Depth: 12.2 m. (40 ft)

PLATE 9. _A tiger rockfish has its dorsal fin raised in defence. This elusive fish will usually dash into rocky crevices when approached. A glass sponge is seen in the background._

Length (fish): 40.6 cm. (16 in.)
Depth: 33.5 m. (110 ft)

PLATE 10. _A quillback rockfish hovers above a cloud sponge. Cloud sponges usually inhabit calm waters in deep fiords. Large masses of these simple creatures offer shelter and protection to fish and invertebrates._

Length (fish): 27.9 cm. (11 in.)
Depth: 36.6 m. (120 ft)

PLATE 11. _Plumose anemones and sea-cucumber. Sea-cucumbers are the vacuum cleaners of the ocean floor. As they inch along the bottom they suck up mud and debris, from which they extract nutrients._

Height (anemones): 66 cm. (26 in.)
Depth: 13.7 m. (45 ft)

PLATE 12. _Flanked by a blood star, a basket starfish has extended its branched arms for feeding. Although radically different in appearance, these animals are related._

Size (basket star): 27.9 cm. (11 in.)
Depth: 13.7 m. (45 ft)

1   BULL KELP

2   DIVERS AT A ROCK-FACE

3   CRIMSON ANEMONE

4 SWIMMING ANEMONE

5   SILVER-TIPPED ANEMONES

6   SCULPIN

7   RED IRISH LORD

8   *TEALIA* ANEMONES

9   TIGER ROCKFISH

10 QUILLBACK ROCKFISH WITH CLOUD SPONGE

11 PLUMOSE ANEMONES AND SEA-CUCUMBER

12　BASKET STARFISH WITH BLOOD STAR

PLATE 13. _Close-up of the mouth and centre portion of a burrowing sea anemone. Food trapped by the tentacles is directed here and ingested._

> _Diameter: 5.1 cm. (2 in.)_
> _Depth: 7.6 m. (25 ft)_

PLATE 14. _Brooding anemones (see also plate 54) cling to branches of calcareous algae._

> _Diameter (anemones): 2.5 cm. (1 in.)_
> _Depth: 9.1 m. (30 ft)_

PLATE 15. _Variegated anemone. This type of anemone is found most often in current-swept areas. Colour varies from white to crimson (see also plates 3 and 16)._

> _Diameter: 20.3 cm. (8 in.)_
> _Depth: 7.6 m. (25 ft)_

PLATE 16. _Close-up of the tentacles and mouth of a crimson anemone. The tentacles are covered with paralyzing cells called nematocysts. When an animal comes in contact with these stinging cells, tiny paralyzing arrows are released into the prey._

> _Size: 7.6 cm. (3 in.)_
> _Depth: 10.7 m. (35 ft)_

PLATE 17. _Tube-worms (see also plate 65) encased in a colonial sea-squirt. This tube-worm lives in a calcareous tube from which it extends a flowery appendage to ensnare tiny organisms. The worm at the top has withdrawn._

> _Height: 5.1 cm. (2 in.)_
> _Depth: 12.2 m. (40 ft)_

PLATE 18. _A blood star feeds on a sponge. Most starfish have voracious appetites and will feed on sponges which are often bypassed by other predators._

> _Size: 12.7 cm. (5 in.)_
> _Depth: 10.7 m. (35 ft)_

PLATE 19. _Sea-pen. This is not a single animal, but rather a colony of creatures sharing the same stalk. When alarmed, the sea-pen will expel water from the central column and deflate into the sand._

> _Height: 45.7 cm. (18 in.)_
> _Depth: 7.6 m. (25 ft)_

PLATE 20. _A Pinto abalone grazes on algae encrusting the rock. Abalone is known not only as a delicious seafood, but also for the mother-of-pearl lining the inside of the shell._

> _Size: 10.2 cm. (4 in.)_
> _Depth: 6.1 m. (20 ft)_

PLATE 21. _Buffalo sculpin. These masters of camouflage are almost impossible to spot when stationary._

> _Length: 20.3 cm. (8 in.)_
> _Depth: 9.1 m. (30 ft)_

PLATE 22. _The Puget Sound king-crab, which in no way resembles the Alaska king-crab, inhabits rocky areas. When harassed it will back into a crevice, wedging itself so firmly that it is all but impossible to dislodge._

> _Size: 20.3 cm. (8 in.)_
> _Depth: 15.2 m. (50 ft)_

PLATE 23. _A rose star marches past a group of solitary orange cup-coral._

> _Size (rose star): 7.6 cm. (3 in.)_
> _Depth: 10.7 m. (35 ft)_

PLATE 24. _Female kelp greenlings line up to dine on a severed sea-urchin._

> _Length (fish): 35.6 cm. (14 in.)_
> _Depth: 13.7 m. (45 ft)_

13   BURROWING SEA-ANEMONE

14   BROODING ANEMONES

15   VARIEGATED ANEMONE

16   CRIMSON ANEMONE

17   TUBE-WORMS IN COLONIAL SEA-SQUIRT

18 BLOOD STAR

19 SEA-PEN

20 PINTO ABALONE

21   BUFFALO SCULPIN

22 PUGET SOUND KING-CRAB

23　ROSE STAR

24   KELP GREENLINGS

PLATE 25. *A snail browses along a kelp stalk.*
Size: 2.5 cm. (1 in.)
Depth: 4.6 m. (15 ft)

PLATE 26. *Tubesnout. This long narrow fish conceals itself by hovering amidst stalks of sea-grass.*
Length: 10.2 cm. (4 in.)
Depth: 4.6 m. (15 ft)

PLATE 27. *Opalescent squid. These molluscs are magnificent swimmers, gracefully and effortlessly propelling themselves through-the water.*
Length: 12.7 cm. (5 in.)
Depth: 3.1 m. (10 ft)

PLATE 28. *Tentacles of a Tealia anemone. The thick stalky tentacles of this species indicate that it has numerous stinging cells and is capable of feeding on larger prey such as shrimp and small fish.*
Length (tentacle): 3.8 cm. (1 1/2 in.)
Depth: 12.2 m. (40 ft)

PLATE 29. *Light-bulb sea-squirts. Each animal draws water into its body through one opening and squirts it out the other after removing nutrients.*
Width: 6.4 cm. (2 1/2 in.)
Depth: 13.7 m. (45 ft)

PLATE 30. *A burrowing sea-cucumber traps organisms on its tentacles. In turn each tentacle is bent into the mouth where the food is swallowed.*
Diameter: 12.7 cm. (5 in.)
Depth: 9.1 m. (30 ft)

PLATE 31. *A sponge has enveloped, then dissolved, the shell of this hermit-crab. In return for providing the crab with protection, the sponge enjoys greater access to food by being carried around.*
Size: 7.6 cm. (3 in.)
Depth: 7.6 m. (25 ft)

PLATE 32. *Slime-star. It will exude thick mucus when attacked. The hole on top is the entrance to the chamber where its young are brooded.*
Size: 10.2 cm. (4 in.)
Depth: 9.1 m. (30 ft)

PLATE 33. *The mouth and radiating tentacles of a burrowing anemone. Burrowing anemones are so named for their ability to withdraw quickly into the muddy bottom when disturbed.*
Size: 3.8 cm. (1 1/2 in.)
Depth: 10.7 m. (35 ft)

PLATE 34. *An anemone engulfs the test, or shell, of a dead giant red sea-urchin.*
Size (anemone): 20.3 cm. (8 in.)
Depth: 7.6 m. (25 ft)

PLATE 35. *Eggs of a sea-slug. Although sea-slugs belong to the mollusc family, which includes clams and snails, like their cousins the octopus and the squid, they have discarded their shells in the course of evolution.*
Size: 7.6 cm. (3 in.)
Depth: 10.7 m. (35 ft)

PLATE 36. *Giant red sea-urchin. The spines are not poisonous, but are quite sharp and painful if stepped or knelt upon. When the animal dies the spines fall off, leaving the skeleton, or test, as seen in plate 34.*
Size: 20.3 cm. (8 in.)
Depth: 7.6 m. (25 ft)

25　SNAIL

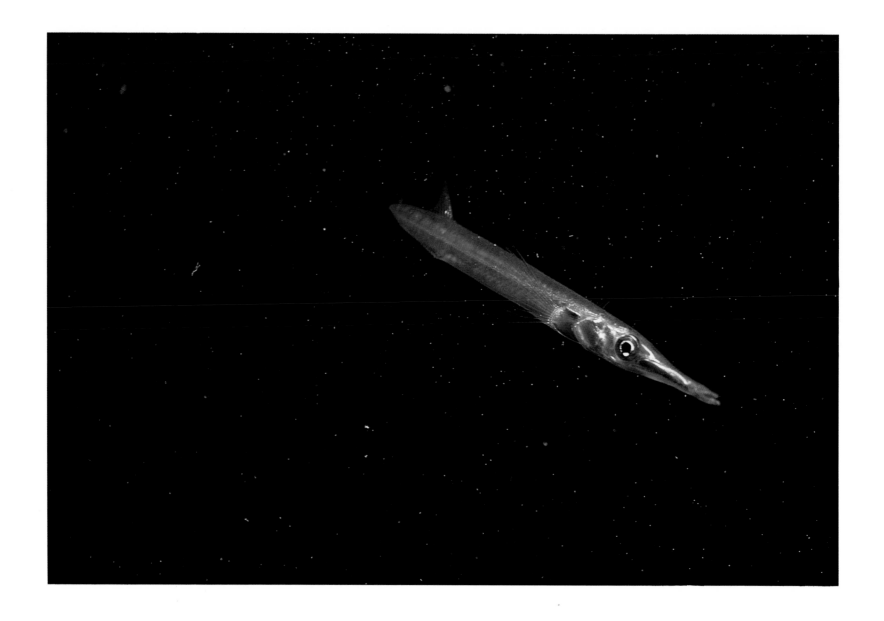

26 TUBESNOUT

27   OPALESCENT SQUID

28 *TEALIA* ANEMONE

29   LIGHT-BULB SEA-SQUIRTS

30 BURROWING SEA-CUCUMBER

31   HERMIT-CRAB WITH SPONGE

32 SLIME-STAR

33   BURROWING ANEMONE

34   ANEMONE WITH SHELL

35   EGGS OF SEA-SLUG

36   GIANT RED SEA-URCHIN

PLATE 37. _Pacific spiny lumpsucker. This fish is so rotund that it makes a poor swimmer. It seems to enjoy moving backwards as much as forwards._

> Length: 5.1 cm. (2 in.)
> Depth: 3.1 m. (10 ft)

PLATE 38. _Male wolf eel. Wolf eels are elongated fish that grow up to 2.4 metres (8 feet) in length. They are thought to mate for life, sharing the same rocky den. His spouse follows in plate 39._

> Length: 1.8 m. (6 ft)
> Depth: 9.1 m. (30 ft)

PLATE 39. _Female wolf eel._

> Length: 1.5 m. (5 ft)
> Depth: 9.1 m. (30 ft)

PLATE 40. _A male lingcod guards his yellowish egg mass. The female departs after laying the eggs, leaving the male to guard the brood. This he does with such ferocious dedication that he will even lunge at a scuba diver who comes too close._

> Length: 40.6 cm. (16 in.)
> Depth: 18.3 m. (60 ft)

PLATE 41. _A male wolf eel seizes at a crab. The powerful jaws easily dispose of crabs—shell and all—and even of spiny sea-urchins._

> Length: 1.2 m. (4 ft)
> Depth: 24.4 m. (80 ft)

PLATE 42. _White-spotted_ Tealia _anemone with tentacles retracted. Only the tips are exposed at the top of the ball._

> Height: 7.6 cm. (3 in.)
> Depth: 10.7 m. (35 ft)

PLATE 43. _The anemone shown in plate 42, with tentacles unfurled._

> Height: 15.2 cm. (6 in.)
> Depth: 10.7 m. (35 ft)

PLATE 44. _Red rock-crab. As a seafood the red rock-crab is not as popular as the larger Dungeness crab but it is every bit as tasty._

> Size: 12.7 cm. (5 in.)
> Depth: 6.1 m. (20 ft)

PLATE 45. _Hermit-crab. Having the eyes located at the ends of stalks permits the hermit-crab to see that all is safe before emerging from the protection of its shell._

> Size: 3.8 cm. (1¹/₂ in.)
> Depth: 4.6 m. (15 ft)

PLATE 46. _White-lined_ Dirona. _Nudibranchs, or sea-slugs, are distant relatives of the common garden slug._

> Length: 12.7 cm. (5 in.)
> Depth: 6.1 m. (20 ft)

PLATE 47. _Transparent sea-squirt (see also plate 29)._

> Height: 3.8 cm. (1¹/₂ in.)
> Depth: 9.1 m. (30 ft)

PLATE 48. _Vermilion starfish. Starfish breathe through their skin. Tiny plier-like organs prevent potential inhabitants from clogging the breathing surface, giving starfish a tidy appearance._

> Size: 15.2 cm. (6 in.)
> Depth: 15.2 m. (50 ft)

37   PACIFIC SPINY LUMPSUCKER

38   MALE WOLF EEL

39   FEMALE WOLF EEL

40 LINGCOD

41 WOLF EEL WITH CRAB

42   *TEALIA* ANEMONE

43 *TEALIA* ANEMONE

44   RED ROCK-CRAB

45   HERMIT-CRAB

46 NUDIBRANCH

47   TRANSPARENT SEA-SQUIRT

48   VERMILION STARFISH

PLATE 49. Steller's sea-lion. These mammals move clumsily on land but exhibit speed and agility underwater. Weighing in at up to 910 kilograms (2000 pounds) and reaching 3.1 metres (10 feet) in length, Steller's sea-lions are to be respected both in and out of the water.

Length: 2.1 m. (7 ft)
Depth: 4.6 m. (15 ft)

PLATE 50. Black rockfish and plumose anemones.
Length (rockfish): 30.5 cm. (1 ft)
Depth: 15.2 m. (50 ft)

PLATE 51. A friendly harbour seal plays tug-of-war with a scuba diver's fin. Harbour seals grow to 1.8 metres (6 feet) in length and are commonly seen hauled out on rocky islets throughout the Pacific northwest.

Length: 1.2 m. (4 ft)
Depth: 6.1 m. (20 ft)

PLATE 52. Cyanea jellyfish with scuba diver in background. The trailing tentacles of these sea blubbers can sting, leaving a painful burning welt.
Diameter: 50.8 cm. (20 in.)
Depth: 18.3 m. (60 ft)

PLATE 53. Pacific spiny dogfish. The dogfish is the most prevalent of sharks in the Pacific northwest. Though the sight of sharks may be fearful, to watch them move effortlessly through the water is to witness the perfection of evolution.

Length: 1.1 m. (3¹/₂ ft)
Depth: 7.6 m. (25 ft)

PLATE 54. Brooding anemone and young. Here a parent, with its tentacles partially withdrawn, is surrounded by its offspring. When mature, the young will leave the parent stalk.
Diameter (parent): 3.8 cm. (1¹/₂ in.)
Depth: 12.2 m. (40 ft)

PLATE 55. Colony of hydroids. Hydroids are members of the coral family and often grow in colonies comprised of many individuals.
Size: 25.4 cm. (10 in.)
Depth: 16.8 m. (55 ft)

PLATE 56. Juvenile yelloweye rockfish. Known also as red snappers, these fish frequent rocky reefs where they can dash into crevices if threatened.
Length: 40.6 cm. (16 in.)
Depth: 19.8 m. (65 ft)

PLATE 57. Red burrowing sea-cucumber.
Height: 15.2 cm. (6 in.)
Depth: 16.8 m. (55 ft)

PLATE 58. Barnacles feed by sweeping their feathery legs through the water. A green sea-urchin perches in the background.
Size (barnacle): 5.1 cm. (2 in.)
Depth: 12.2 m. (40 ft)

PLATE 59. View of a sea-slug from the rear.
Length: 3.8 cm. (1¹/₂ in.)
Depth: 7.6 m. (25 ft)

PLATE 60. Bull kelp frond dangling in water.
Length: 2.4 m. (8 ft)
Depth: 4.6 m. (15 ft)

49 STELLER'S SEA-LION

50   BLACK ROCKFISH

51   HARBOUR SEAL

52 *CYANEA* JELLYFISH

53   PACIFIC SPINY DOGFISH

54 BROODING ANEMONE

55 HYDROIDS

56 YELLOWEYE ROCKFISH

57   SEA-CUCUMBER

58 BARNACLES

59 SEA-SLUG

60 BULL KELP

PLATE 61. *Young bull kelp. In its quest for the surface and the life-giving sun, it will grow up to 30 centimetres (1 foot) per day.*
Height: 76 cm. (2$^1/_2$ ft)
Depth: 7.6 m. (25 ft)

PLATE 62. *Giant Pacific octopus and scuba diver. Octopuses are generally shy creatures that must be coaxed out of their rocky dens. Proper care in handling them ensures that their unprotected organs will not be damaged.*
Size (octopus): 1.8 m. (6 ft)
Depth: 9.1 m. (30 ft)

PLATE 63. *The giant Pacific octopus, largest of all living octopuses, has the ability to change colours. The red hue of this octopus may indicate aggression.*
Size: 1.5 m. (5 ft)
Depth: 7.6 m. (25 ft)

PLATE 64. *Swimming scallop coated with rough encrusting sponge. These scallops are able to propel themselves erratically through the water by rapidly opening and closing their hinged shells. The eyes are clearly visible in this photograph.*
Size: 5.1 cm. (2 in.)
Depth: 9.1 m. (30 ft)

PLATE 65. *The tube-worm traps its food upon a fine network of bristles. When these bristles are withdrawn the stopper seen in this photograph blocks the entrance to the tube, thereby protecting the animal.*
Size: 2.5 cm. (1 in.)
Depth: 6.1 m. (20 ft)

PLATE 66. *Yellow boring sponge. Boring sponges chemically decompose the shells of molluscs. Scallop shells are the preferred victims of the yellow boring sponge.*
Width: 3.8 cm. (1$^1/_2$ in.)
Depth: 15.2 m. (50 ft)

PLATE 67. *A quillback rockfish seeks refuge in the hollow of a cloud sponge.*
Length: (rockfish): 22.9 cm. (9 in.)
Depth: 36.6 m. (120 ft)

PLATE 68. *An orange-peel nudibranch inches over the holdfasts of kelp.*
Size: 12.7 cm. (5 in.)
Depth: 7.6 m. (25 ft)

PLATE 69. *Purple sea-star and anemones. This species is the most commonly encountered starfish, often exposed on a rocky shore at low tide.*
Size (starfish): 22.9 cm. (9 in.)
Depth: 12.2 m. (40 ft)

PLATE 70. *Hydroids (see also plate 55).*
Height: 3.8 cm. (1$^1/_2$ in.)
Depth: 10.7 m. (35 ft)

PLATE 71. *Starry flounder on sandy bottom. Flounders begin life oriented like most fish, but as they mature they turn on one side and assume a bottom-dwelling existence. Flounders are capable of altering their colour patterns to mimic the bottom upon which they rest.*
Length: 35.6 cm. (14 in.)
Depth: 6.1 m. (20 ft)

PLATE 72. *Common Aeolid nudibranch.*
Length: 2.5 cm. (1 in.)
Depth: 13.7 m. (45 ft)

61 YOUNG BULL KELP

62   GIANT PACIFIC OCTOPUS

63   GIANT PACIFIC OCTOPUS

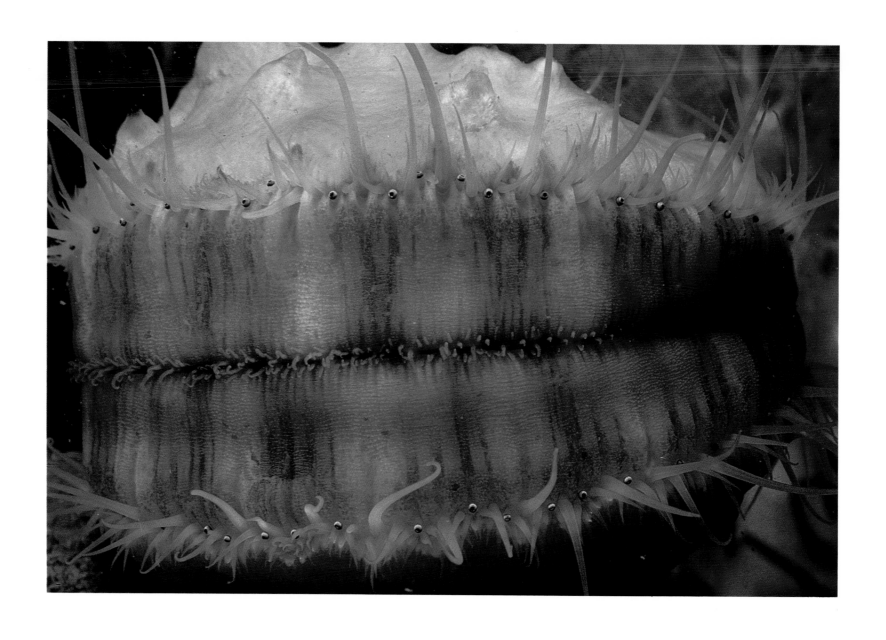

64   SWIMMING SCALLOP

65 TUBE-WORM

66 YELLOW BORING SPONGE

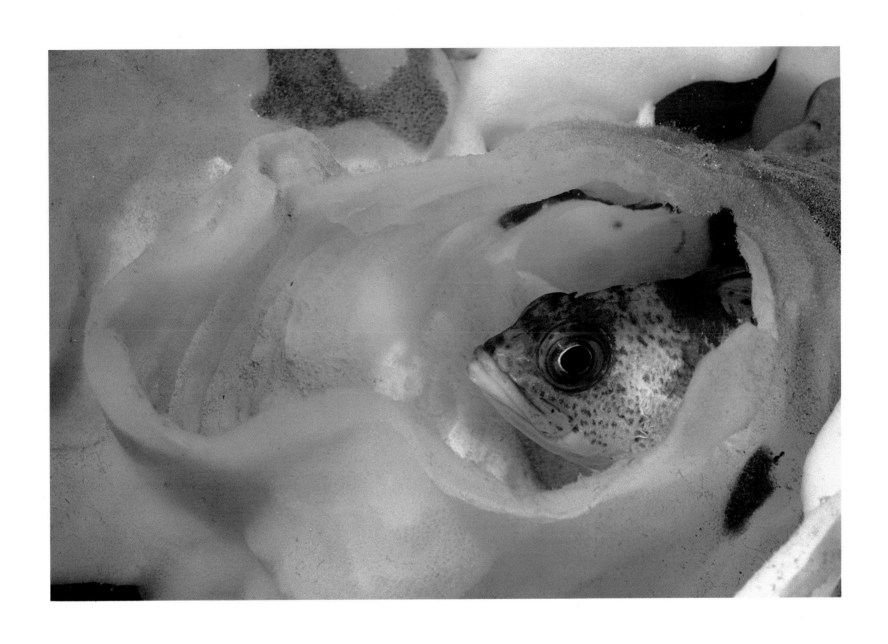

67 QUILLBACK ROCKFISH

68   NUDIBRANCH

69   PURPLE SEA-STAR WITH ANEMONES

70 HYDROIDS

71 STARRY FLOUNDER

72 NUDIBRANCH

PLATE 73. Tealia _anemone. This sea-anemone feeds on unsuspecting small fish._
Height: 15.2 cm. (6 in.)
Depth: 9.1 m. (30 ft)

PLATE 74. China rockfish.
Length: 35.6 cm. (14 in.)
Depth: 24.4 m. (80 ft)

PLATE 75. Lined chiton. Chitons generally have eight overlapping plates that give structural rigidity to the body. These relatives of the snail feed on vegetation by scraping it off rocks.
Length: 3.8 cm. (1½ in.)
Depth: 9.1 m. (30 ft)

PLATE 76. Two colour variations of the short-spined starfish.
Size: 25.4 cm. (10 in.)
Depth: 4.6 m. (15 ft)

PLATE 77. Variegated anemone.
Height: 15.2 cm. (6 in.)
Depth: 7.6 m. (25 ft)

PLATE 78. A hermit-crab searches for an empty snail shell. This family of crabs, having soft abdomens, must reside in discarded shells for protection.
Size: 1.9 cm. (¾ in.)
Depth: 10.7 m. (35 ft)

PLATE 79. Hydroids and orange cup-coral.
Height: (hydroids): 22.9 cm. (9 in.)
Depth: 12.2 m. (40 ft)

PLATE 80. A kelp crab dines on a piece of algae.
Size: 7.6 cm. (3 in.)
Depth: 4.6 m. (15 ft)

PLATE 81. Decorated warbonnet. The bizarre ornamentation of the head effectively camouflages this long slender fish in its preferred habitat among hydroids, sponges, and algae.
Length: 17.8 cm. (7 in.)
Depth: 16.8 m. (55 ft)

PLATE 82. Coralline alga. The calcified segments of this plant give it structural rigidity.
Height: 7.6 cm. (3 in.)
Depth: 6.1 m. (20 ft)

PLATE 83. Hydroids (see also plates 55 and 70).
Width: 3.8 cm. (1½ in.)
Depth: 12.2 m. (40 ft)

PLATE 84. Strawberry anemones cloak a giant barnacle.
Diameter (anemone): 1.9 cm. (¾ in.)
Depth: 13.7 m. (45 ft)

73 *TEALIA* ANEMONE

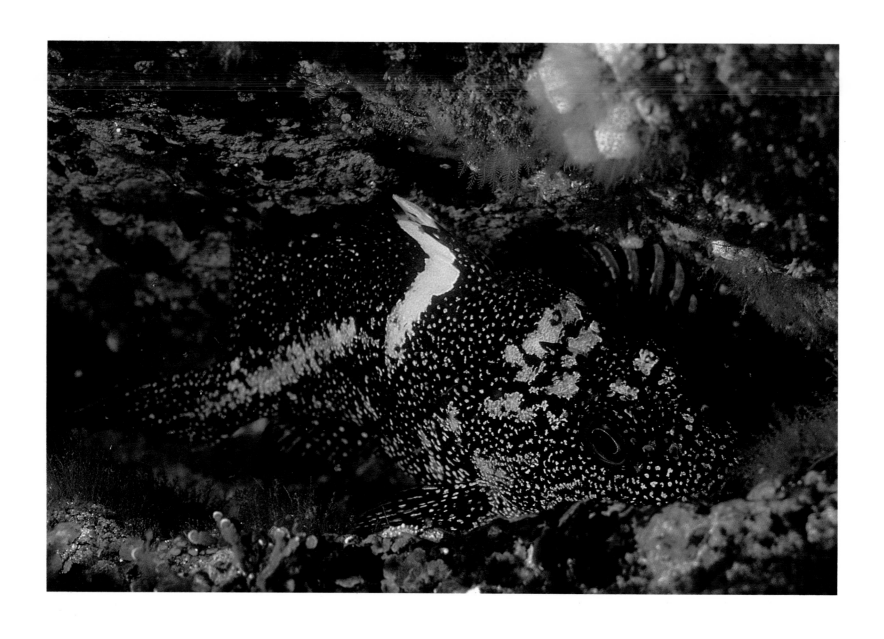

74   CHINA ROCKFISH

75   LINED CHITON

76 SHORT-SPINED STARFISH

77 VARIEGATED ANEMONE

78   HERMIT-CRAB

79 HYDROIDS AND CORAL

80   KELP CRAB

81   DECORATED WARBONNET

82 CORALLINE ALGA

83 HYDROIDS

84   STRAWBERRY ANEMONES

PLATE 85. Branching hydrocoral interlaced with the slender arms of brittle star. A sculpin lurks beneath the white sea-anemones.
>> _Height: (hydrocoral): 7.6 cm. (3 in.)_
>> _Depth: 6.1 m. (20 ft)_

PLATE 86. White sea-cucumber surrounded by colonial sea-squirt.
>> _Height: 7.6 cm. (3 in.)_
>> _Depth: 19.8 m. (65 ft)_

PLATE 87. Sea-squirts. Although primitive in appearance, they are actually members of the group of animals called Chordata which includes man.
>> _Height: 2.5 cm. (1 in.)_
>> _Depth: 15.2 m. (50 ft)_

PLATE 88. California mussels.
>> _Length: 17.8 cm. (7 in.)_
>> _Depth: 9.1 m. (30 ft)_

PLATE 89. Scuba diver with sunflower star (yellow) and striped sunstar (blue colour variation). The sunflower star is the largest species of starfish in the world and may grow to 0.9 metres (3 feet) across.
>> _Depth: 6.1 m. (20 ft)_

PLATE 90. Underside of a striped sunstar, showing stomach. Starfish can eat prey externally by pushing out their stomachs, engulfing the quarry, and digesting it.
>> _Size: 40.6 cm. (16 in.)_
>> _Depth: 18.3 m. (60 ft)_

PLATE 91. Three colour phases of the purple sea-star; sea lettuce; giant red sea-urchin.
>> _Size (urchin): 15.2 cm. (6 in.)_
>> _Depth: 3.1 m. (10 ft)_

PLATE 92. Orange social sea-squirts.
>> _Size (patch): 10.2 cm. (4 in.)_
>> _Depth: 12.2 m. (40 ft)_

PLATE 93. Coon-striped shrimp. This species of shrimp actively prevents organisms from settling on its body, a characteristic also exhibited to varying degrees by other crustaceans such as lobsters and crabs.
>> _Size: 7.6 cm. (3 in.)_
>> _Depth: 6.1 m. (20 ft)_

PLATE 94. Gumboot chiton. This is the largest species of chiton in the world.
>> _Size: 25.4 cm. (10 in.)_
>> _Depth: 4.6 m. (15 ft)_

PLATE 95. Bull kelp and vermilion starfish.
>> _Size (starfish): 17.8 cm. (7 in.)_
>> _Depth: 6.1 m. (20 ft)_

PLATE 96. Bull kelp waving in the current.
>> _Length (blades): 0.9 m. (3 ft)_
>> _Depth: 6.1 m. (20 ft)_

85 HYDROCORAL

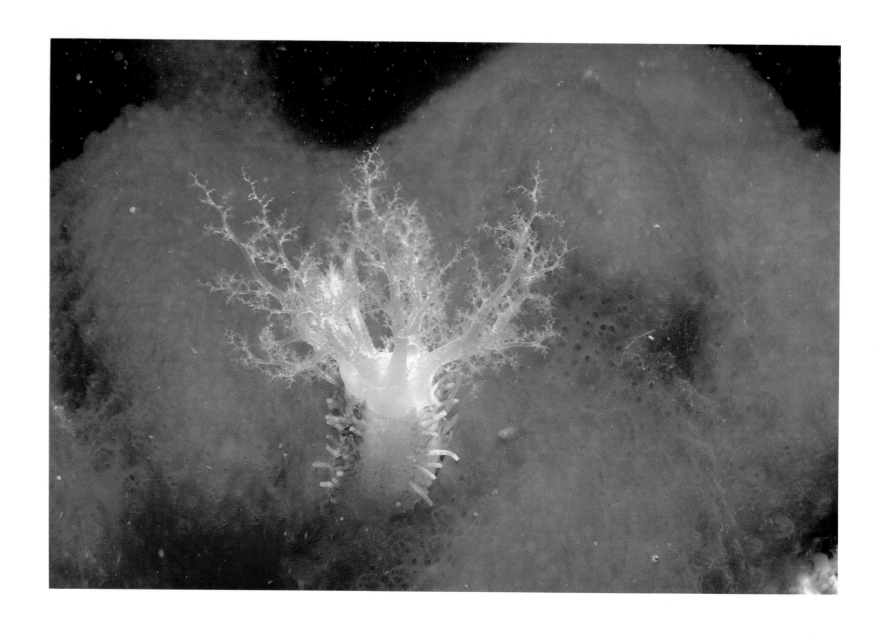

86   WHITE SEA-CUCUMBER

87   SEA-SQUIRTS

88 CALIFORNIA MUSSELS

89 SUNFLOWER STAR AND STRIPED SUNSTAR

90 STRIPED SUNSTAR

91   PURPLE SEA-STARS

92   SEA-SQUIRTS

93   SHRIMP

94   GUMBOOT CHITON

95   BULL KELP WITH VERMILION STARFISH

96 BULL KELP

Photographic Information

PLATE 1: Kodachrome 64; natural light; Nikkor 16mm. f/2.8 lens in Oceanic Hydro 35 housing; exposure 1/60 sec. at f/11.

PLATE 2: Kodachrome 64; two strobes; Nikkor 18mm. f/4 lens in Oceanic Hydro 35 housing; exposure 1/60 sec. at f/5.6.

PLATE 3: Kodachrome 64; natural light with fill-in flash from two strobes; Nikkor 16mm. f/2.8 lens in Oceanic Hydro 35 housing; exposure 1/15 sec. at f/16.

PLATE 4: Kodachrome 64; one strobe; Nikkor 55mm. f/3.5 micro lens in Aquatica housing; exposure 1/60 sec. at f/11.

PLATE 5: Ektachrome 64; one strobe; Nikkor 55mm. f/3.5 micro lens in Aquatica housing; exposure 1/60 sec. at f/11.

PLATE 6: Kodachrome 64; one strobe; Nikkor 55mm. f/3.5 micro lens in Aquatica housing; exposure 1/60 sec. at f/16.

PLATE 7: Kodachrome 64; one strobe; Nikkor 55mm. f/2.8 micro lens in Aquatica housing; exposure 1/60 sec. at f/11.

PLATE 8: Kodachrome 64; one strobe; Nikkor 55mm. f/2.8 micro lens in Aquatica housing; exposure 1/60 sec. at f/11.

PLATE 9: Kodachrome 64; one strobe; Nikkor 55mm. f/2.8 micro lens in Aquatica housing; exposure 1/60 sec. at f/8.

PLATE 10: Kodachrome 64; one strobe; Nikkor 55mm. f/2.8 micro lens in Aquatica housing; exposure 1/60 sec. at f/8.

PLATE 11: Kodachrome 64; two strobes; Nikkor 20mm. f/3.5 lens in Aquatica housing; exposure 1/60 sec. at f/11.

PLATE 12: Kodachrome 64; one strobe; Nikkor 55mm. f/3.5 micro lens in Aquatica housing; exposure 1/60 sec. at f/11.

PLATE 13: Kodachrome 64; one strobe; Nikkor 55mm. f/3.5 micro lens with PK-13 extension ring in Ikelite housing; exposure 1/60 sec. at f/22.

PLATE 14: Kodachrome 64; one strobe; Nikkor 55mm. f/3.5 micro lens in Aquatica housing; exposure 1/60 sec. at f/16.

PLATE 15: Kodachrome 64; one strobe; Nikkor 55mm. f/3.5 micro lens in Aquatica housing; exposure 1/60 sec. at f/11.

PLATE 16: Kodachrome 64; one strobe; Nikkor 55mm. f/3.5 micro lens with PK-13 extension ring in Ikelite housing; exposure 1/60 sec. at f/22.

PLATE 17: Kodachrome 64; one strobe; Nikkor 55mm. f/3.5 micro lens in Aquatica housing; exposure 1/60 sec. at f/16.

PLATE 18: Kodachrome 64; one strobe; Nikkor 55mm. f/2.8 micro lens in Aquatica housing; exposure 1/60 sec. at f/11.

PLATE 19: Kodachrome 64; one strobe; Nikkor 55mm. f/3.5 micro lens in Aquatica housing; exposure 1/60 sec. at f/11.

PLATE 20: Ektachrome 64; one strobe; Nikkor 55mm. f/3.5 micro lens in Aquatica housing; exposure 1/60 sec. at f/8.

PLATE 21: Ektachrome 64; one strobe; Nikkor 55mm. f/3.5 micro lens in Aquatica housing; exposure 1/60 sec. at f/11.

PLATE 22: Ektachrome 64; one strobe; Nikkor 55mm. f/3.5 micro lens in Aquatica housing; exposure 1/60 sec. at f/11.

PLATE 23: Kodachrome 64; one strobe; Nikkor 55mm. f/3.5 micro lens in Aquatica housing; exposure 1/60 sec. at f/11.

PLATE 24: Kodachrome 64; two strobes; Nikkor 24mm. f/2 lens in Aquatica housing; exposure 1/30 sec. at f/8.

PLATE 25: Kodachrome 64; one strobe; Nikkor 55mm. f/2.8 micro lens in Aquatica housing; exposure 1/60 sec. at f/16.

PLATE 26: Kodachrome 64; one strobe; Nikkor 55mm. f/3.5 micro lens with PK-13 extension ring in Ikelite housing; exposure 1/60 sec. at f/11.

PLATE 27: Kodachrome 64; one strobe; Nikkor 55mm. f/3.5 micro lens in Aquatica housing; exposure 1/60 sec. at f/11.

PLATE 28: Kodachrome 64; one strobe; Nikkor 55mm. f/3.5 micro lens in Aquatica housing; exposure 1/60 sec. at f/16.

PLATE 29: Kodachrome 64; one strobe; Nikkor 55mm. f/3.5 micro lens with PK-13 extension ring in Ikelite housing; exposure 1/60 sec. at f/16.

PLATE 30: Kodachrome 64; one strobe; Nikkor 55mm. f/3.5 micro lens with PK-13 extension ring in Ikelite housing; exposure 1/60 sec. at f/22.

PLATE 31: Kodachrome 64; one strobe; Nikkor 55mm. f/3.5 micro lens in Aquatica housing; exposure 1/60 sec. at f/16.

PLATE 32: Ektachrome 64; one strobe; Nikkor 55mm. f/3.5 micro lens in Aquatica housing; exposure 1/60 sec. at f/11.

PLATE 33: Ektachrome 64; one strobe; Nikkor 55mm. f/3.5 micro lens with PK-13 extension ring in Ikelite housing; exposure 1/60 sec. at f/22.

PLATE 34: Kodachrome 64; one strobe; Nikkor 55mm. f/2.8 micro lens in Aquatica housing; exposure 1/60 sec. at f/11.

PLATE 35: Kodachrome 64; one strobe; Nikkor 55mm. f/3.5 micro lens with PK-13 extension ring in Ikelite housing; exposure 1/60 sec. at f/16.

PLATE 36: Kodachrome 64; one strobe; Nikkor 55mm. f/3.5 micro lens in Aquatica housing; exposure 1/60 sec. at f/8.

PLATE 37: Ektachrome 64; one strobe; Nikkor 55mm. f/3.5 micro lens with PK-13 extension ring in Ikelite housing; exposure 1/60 sec. at f/16.

PLATE 38: Kodachrome 64; one strobe; Nikkor 55mm. f/3.5 micro lens in Aquatica housing; exposure 1/60 sec. at f/16.

PLATE 39: Kodachrome 64; one strobe; Nikkor 55mm. f/3.5 micro lens in Aquatica housing; exposure 1/60 sec. at f/11.

PLATE 40: Kodachrome 64; one strobe; Nikkor 55mm. f/3.5 micro lens in Aquatica housing; exposure 1/60 sec. at f/8.

PLATE 41: Kodachrome 64; one strobe; Nikkor 55mm. f/3.5 micro lens in Aquatica housing; exposure 1/60 sec. at f/11.

PLATE 42: Kodachrome 64; one strobe; Nikkor 55mm. f/3.5 micro lens in Aquatica housing; exposure 1/60 sec. at f/16.

PLATE 43: Kodachrome 64; one strobe; Nikkor 55mm. f/3.5 micro lens in Aquatica housing; exposure 1/60 sec. at f/11.

PLATE 44: Kodachrome 64; one strobe; Nikkor 55mm. f/3.5 micro lens in Aquatica housing; exposure 1/60 sec. at f/16.

PLATE 45: Kodachrome 64; one strobe; Nikkor 55mm. f/3.5 micro lens with PK-13 extension ring in Ikelite housing; exposure 1/60 sec. at f/16.

PLATE 46: Kodachrome 64; one strobe; Nikkor 55mm. f/3.5 micro lens in Aquatica housing; exposure 1/60 sec. at f/16.

PLATE 47: Ektachrome 64; one strobe; Nikkor 55mm. f/3.5 micro lens with PK-13 extension ring in Ikelite housing; exposure 1/60 sec. at f/22.

PLATE 48: Kodachrome 64; one strobe; Nikkor 55mm. f/3.5 micro lens in Aquatica housing; exposure 1/60 sec. at f/8.

PLATE 49: Ektachrome 64; one strobe with natural light; Nikkor 24mm. f/2 lens in Aquatica housing; exposure 1/60 sec. at f/8.

PLATE 50: Kodachrome 64; two strobes; Nikkor 20mm. f/3.5 lens in Aquatica housing; exposure 1/60 sec. at f/5.6.

PLATE 51: Ektachrome 64; one strobe; Nikkor 24mm. f/2 lens in Aquatica housing; exposure 1/60 sec. at f/5.6.

PLATE 52: Kodachrome 64; one strobe; Nikkor 24mm. f/2 lens in Aquatica housing; exposure 1/15 sec. at f/11.

PLATE 53: Kodachrome 64; two strobes; Nikkor 24mm. f/2 lens in Aquatica housing; exposure 1/60 sec. at f/8.

PLATE 54: Kodachrome 64; one strobe; Nikkor 55mm. f/3.5 micro lens with PK-13 extension ring in Ikelite housing; exposure 1/60 sec. at f/16.

PLATE 55: Kodachrome 64; one strobe; Nikkor 55mm. f/3.5 micro lens in Aquatica housing; exposure 1/60 sec. at f/8.

PLATE 56: Kodachrome 64; one strobe; Nikkor 55mm. f/3.5 micro lens in Aquatica housing; exposure 1/60 sec. at f/8.

PLATE 57: Kodachrome 64; one strobe; Nikkor 55mm. f/2.8 micro lens in Aquatica housing; exposure 1/60 sec. at f/11.

PLATE 58: Ektachrome 64; one strobe; Nikkor 55mm. f/3.5 micro lens in Aquatica housing; exposure 1/60 sec. at f/16.

PLATE 59: Ektachrome 64; one strobe; Nikkor 55mm. f/3.5 micro lens with PK-13 extension ring in Ikelite housing; exposure 1/60 sec. at f/22.

PLATE 60: Kodachrome 64; natural light; Nikkor 24mm. f/2 lens in Aquatica housing; exposure 1/125 sec. at f/8.

PLATE 61: Kodachrome 64; two strobes with natural light; Nikkor 20mm. f/3.5 lens in Aquatica housing; exposure 1/60 sec. at f/5.6.

PLATE 62: Kodachrome 64; two strobes with natural light; Nikkor 16mm. f/2.8 lens in Oceanic Hydro 35 housing; exposure 1/30 sec. at f/11.

PLATE 63: Kodachrome 64; two strobes with natural light; Nikkor 16mm. f/2.8 lens in Oceanic Hydro 35 housing; exposure 1/30 sec. at f/16.

PLATE 64: Kodachrome 64; one strobe; Nikkor 55mm. f/3.5 micro lens with PK-13 extension ring in Ikelite housing; exposure 1/60 sec. at f/16.

PLATE 65: Ektachrome 64; one strobe; Nikkor 55mm. f/3.5 micro lens with PK-13 extension ring in Ikelite housing; exposure 1/60 sec. at f/16.

PLATE 66: Kodachrome 64; one strobe; Nikkor 55mm. f/3.5 micro lens with PK-13 extension ring in Ikelite housing; exposure 1/60 sec. at f/16.

PLATE 67: Kodachrome 64; one strobe; Nikkor 55mm. f/2.8 micro lens in Aquatica housing; exposure 1/60 sec. at f/8.

PLATE 68: Kodachrome 64; one strobe; Nikkor 55mm. f/2.8 micro lens in Aquatica housing; exposure 1/60 sec. at f/11.

PLATE 69: Kodachrome 64; one strobe; Nikkor 55mm. f/2.8 micro lens in Aquatica housing; exposure 1/60 sec. at f/8.

PLATE 70: Ektachrome 64; one strobe; Nikonos with 35mm. lens and 1:1 extension tube; exposure 1/60 sec. at f/11.

PLATE 71: Ektachrome 64; one strobe; Nikkor 55mm. f/3.5 micro lens in Aquatica housing; exposure 1/60 sec. at f/8.

PLATE 72: Kodachrome 64; one strobe; Nikkor 55mm. f/3.5 micro lens with PK-13 extension ring in Ikelite housing; exposure 1/60 sec. at f/16.

PLATE 73: Kodachrome 64; one strobe; Nikkor 55mm. f/3.5 micro lens in Aquatica housing; exposure 1/60 sec. at f/8.

PLATE 74: Kodachrome 64; one strobe; Nikkor 55mm. f/3.5 micro lens in Aquatica housing; exposure 1/60 sec. at f/8.

PLATE 75: Ektachrome 64; one strobe; Nikkor 55mm. f/3.5 micro lens with PK-13 extension ring in Ikelite housing; exposure 1/60 sec. at f/16.

PLATE 76: Ektachrome 64; one strobe; Nikkor 55mm. f/3.5 micro lens in Aquatica housing; exposure 1/60 sec. at f/11.

PLATE 77: Kodachrome 64; one strobe; Nikkor 55mm. f/3.5 micro lens in Aquatica housing; exposure 1/60 sec. at f/11.

PLATE 78: Kodachrome 64; one strobe; Nikkor 55mm. f/3.5 micro lens with PK-13 extension ring in Ikelite housing; exposure 1/60 sec. at f/11.

PLATE 79: Ektachrome 64; one strobe; Nikkor 55mm. f/3.5 micro lens in Aquatica housing; exposure 1/60 sec. at f/11.

PLATE 80: Kodachrome 64; one strobe; Nikkor 55mm. f/3.5 micro lens in Aquatica housing; exposure 1/60 sec. at f/11.

PLATE 81: Kodachrome 64; one strobe; Nikkor 55mm. f/3.5 micro lens in Aquatica housing; exposure 1/60 sec. at f/16.

PLATE 82: Kodachrome 64; one strobe; Nikkor 55mm. f/3.5 micro lens in Aquatica housing; exposure 1/60 sec. at f/11.

PLATE 83: Kodachrome 64; one strobe; Nikkor 55mm. f/3.5 micro lens with PK-13 extension ring in Ikelite housing; exposure 1/60 sec. at f/22.

PLATE 84: Kodachrome 64; one strobe; Nikkor 55mm. f/2.8 micro lens in Aquatica housing; exposure 1/60 sec. at f/16.

PLATE 85: Kodachrome 64; one strobe; Nikkor 55mm. f/3.5 micro lens in Aquatica housing; exposure 1/60 sec. at f/16.

PLATE 86: Kodachrome 64; one strobe; Nikkor 55mm. f/3.5 micro lens in Aquatica housing; exposure 1/60 sec. at f/11.

PLATE 87: Kodachrome 64; one strobe; Nikkor 55mm. f/3.5 micro lens with PK-13 extension ring in Ikelite housing; exposure 1/60 sec. at f/16.

PLATE 88: Kodachrome 64; one strobe; Nikkor 55mm. f/2.8 micro lens in Aquatica housing; exposure 1/60 sec. at f/11.

PLATE 89: Kodachrome 64; natural light; Nikkor 20mm. f/3.5 lens in Aquatica housing; exposure 1/30 sec. at f/5.6.

PLATE 90: Ektachrome 64; one strobe; Nikkor 55mm. f/3.5 micro lens in Aquatica housing; exposure 1/60 sec. at f/8.

PLATE 91: Ektachrome 64; one strobe with natural light; Nikkor 20mm. f/3.5 lens in Aquatica housing; exposure 1/60 sec. at f/5.6.

PLATE 92: Kodachrome 64; one strobe; Nikkor 55mm. f/3.5 micro lens with PK-13 extension ring in Ikelite housing; exposure 1/60 sec. at f/16.

PLATE 93: Kodachrome 64; one strobe; Nikkor 55mm. f/3.5 micro lens in Aquatica housing; exposure 1/60 sec. at f/16.

PLATE 94: Kodachrome 64; one strobe; Nikkor 55mm. f/3.5 micro lens in Aquatica housing; exposure 1/60 sec. at f/8.

PLATE 95: Kodachrome 64; one strobe with natural light; Nikkor 16mm. f/2.8 lens in Oceanic Hydro 35 housing; exposure 1/60 sec. at f/11.

PLATE 96: Kodachrome 64; natural light; Nikkor 18mm. f/4 lens in Oceanic Hydro 35 housing; exposure 1/125 sec. at f/8.